A book presented by:
sunwave4u@gmail.com

LIVING BY SCIENCE

A Practical Guide to Scientific Living

Inderbir Singh

Knowledge is Limitless

All Rights Reserved © 2024 – Inderbir Singh

All rights reserved. No part of this book may be reproduced, stored in a retrieval system, or transmitted in any form or by any means, electronic, mechanical, photocopying, recording, or otherwise, without the prior written permission of the author.

This book is a work of nonfiction. While the author has made every effort to ensure the accuracy and completeness of the information contained within, the author assumes no responsibility for errors, inaccuracies, omissions, or any consequences arising from the use of this material.

Any trademarks, service marks, product names, or named features are assumed to be the property of their respective owners and are used only for reference. There is no implied endorsement if any of these terms are used.

For permission requests, please contact the author at Email:
sunwave4u@gmail.com.

© [2024] [Inderbir Singh]

- - - - - o - - - - -

DISCLAIMER

The information provided in this book is intended for educational and informational purposes only. It is not a substitute for professional advice, diagnosis, or treatment. Always seek the advice of your physician or other qualified health providers with any questions you may have regarding a medical condition or mental health concerns.

The practices and techniques discussed in this book are meant to support personal growth and well-being. Individual experiences may vary, and results are not guaranteed. Use your own judgment and intuition when applying the information provided.

The author and publisher disclaim any liability for any direct, indirect, or consequential loss or damage arising from the use or reliance on the material contained in this book. Always take appropriate precautions and consult with a professional before beginning any new health or wellness practice.

By using this book, you acknowledge that you are responsible for your own health and well-being and that the author and publisher are not responsible for any outcomes resulting from the application of the information provided.

Thank you for your understanding.

DEDICATION

To my family, whose unwavering support and encouragement have been the foundation of my journey.

To the scientists, educators, and thinkers who dedicate their lives to the pursuit of knowledge and the betterment of humanity.

And to all those who strive to live thoughtfully, rationally, and scientifically – may this book serve as a guide and inspiration on your path.

This book is for you.
Inderbir Singh

Published by Inderbir Singh
Printed in the United States of America
A book by Inderbir Singh (sunwave4u@gmail.com)
Available from Amazon.com and other retail outlets

First Printing Edition, 2024

Contents

Preface 5
Acknowledgment 7

Part 1: Understanding Science 10

Chapter 1: The Scientific Method 11

An Overview Of The Scientific Method & How It Can Be Applied To Everyday Life 12
Hypotheses, Experiments, Data Analysis & Conclusions 16

Chapter 2: Critical Thinking 20

Developing Critical Thinking Skills To Evaluate Information & Make Informed Decisions 21
Recognizing Biases, Fallacies & Logical Errors 26

Part 2: Scientific Nutrition 30

Chapter 3: The Science of Nutrition 31

The Fundamentals Of Nutrition & The Role Of Macronutrients & Micronutrients In The Body 32

Debunking Common Myths & Misconceptions About Diet & Nutrition 38

Chapter 4: Building a Healthy Diet 42

Practical Tips For Designing A Balanced & Nutritious Diet Based On Scientific Evidence... 43

Understanding Portion Control, Meal Planning & Mindful Eating 48

Part 3: Fitness & Exercise 52

Chapter 5: Exercise Physiology... 53

Understanding How Exercise Affects The Body At The Cellular & Molecular Levels 54

Exploring Different Types Of Exercise & Their Benefits 59

Chapter 6: Designing an Effective Workout Routine 65

Guidelines For Creating A Personalized Workout Routine That Aligns With Your Fitness Goals...................................... 66

Incorporating Strength Training, Cardiovascular Exercise, Flexibility & Recovery 72

Part 4: Mental Health & Well-being... 77

sunwave4u@gmail.com

Chapter 7: The Science of Happiness 78

    Exploring The Psychology Of Happiness & The Factors That Contribute To Well-Being 79

    Strategies For Cultivating Positive Emotions & Resilience ... 85

Chapter 8: Stress Management 90

    Understanding The Physiological & Psychological Effects Of Stress On The Body 91

    Techniques For Managing Stress & Promoting Relaxation, Including Mindfulness & Meditation 96

Part 5: Environmental Sustainability 105

Chapter 9: Sustainable Living 106

    The Importance Of Environmental Sustainability For Personal & Planetary Health 107

    Practical Tips For Reducing Your Carbon Footprint & Living More Sustainably 112

Chapter 10: Consumer Choices 118

    Making Environmentally Conscious Choices When Purchasing Food, Clothing, Household Products & Electronics 119

sunwave4u@gmail.com

Evaluating Eco-Friendly Certifications & Labels................ 125

Conclusion: **130**

About the Author: 131

sunwave4u@gmail.com

Preface

In an era marked by rapid technological advancements and unprecedented access to information, the ability to think scientifically and make evidence-based decisions is more critical than ever. Yet, amidst the constant flow of data and the complexities of modern life, many people find it challenging to integrate scientific principles into their daily routines. This book, "Living by Science: A Practical Guide to Scientific Living," aims to bridge that gap, offering readers practical strategies for applying the scientific method to everyday life.

"Living by Science" is not just about understanding the world through a scientific lens; it's about making informed choices that improve our health, enhance our well-being, and foster a deeper connection to the natural world. From the food we eat and the exercises we perform to the way we manage stress and make sustainable choices, this book provides a roadmap for living thoughtfully and rationally.

Drawing on a wealth of scientific research and real-world

examples, this book covers a wide range of topics, including critical thinking, nutrition, exercise, mental well-being, and environmental sustainability. Each chapter is designed to be both informative and actionable, offering insights that can be immediately applied to enhance your daily life.

I have always been passionate about the power of science to improve our lives and the world around us. My hope is that this book will empower you to harness that power, encouraging you to ask questions, seek evidence, and make decisions based on sound scientific principles. Whether you are a seasoned science enthusiast or someone just beginning to explore the world of evidence-based living, this book is for you.

Thank you for embarking on this journey with me. Together, we can cultivate a life that is not only more informed and intentional but also more fulfilling and sustainable.

Warm regards,

INDERBIR SINGH
sunwave4u@gmail.com

Acknowledgment

I would like to express my deepest gratitude to all those who have contributed to the creation of this book "Living by Science: A Practical Guide to Scientific Living".

First and foremost, I am immensely thankful to my family for their unwavering love, support, and encouragement throughout this journey. Their patience and understanding have been a constant source of strength and inspiration.

Special thanks to internet & social media which have provided immense knowledge to shape up this book.

I am deeply grateful to the teachers, mentors, and authors whose wisdom and teachings have significantly influenced our understanding of living by science. Their work has been a guiding light and source of inspiration.

A special thanks to my friends and colleagues for their encouragement, feedback, and support throughout the writing

sunwave4u@gmail.com

process. Your encouragement and enthusiasm have kept me motivated and inspired to bring this project to fruition.

Last but not least, I would like to express my gratitude to the readers who have embarked on this journey with me. It is my sincere hope that this book will offer valuable insights, inspiration, and guidance on the path to deeper understanding and fulfillment in love.

Thank you all for being a part of this incredible journey.

With heartfelt gratitude,
INDERBIR SINGH

sunwave4u@gmail.com

INDERBIR SINGH
Knowledge is Limitless

sunwave4u@gmail.com

Part 1: Understanding Science

1.1: The Scientific Method

An Overview Of The Scientific Method & How It Can Be Applied To Everyday Life.

The scientific method is a systematic approach to inquiry that involves making observations, forming hypotheses, conducting experiments, analyzing data, and drawing conclusions. While traditionally associated with scientific research, the principles of the scientific method can also be applied to various aspects of everyday life to make informed decisions and solve problems effectively.

1. Observation:

- ❖ The scientific method begins with observation, where one notices a phenomenon or pattern in the world around them.

- ❖ In daily life, observation might involve noticing changes in behavior, environment, or physical sensations.

2. Formulation of a Hypothesis:

- ❖ Based on observations, a hypothesis is formulated as a tentative explanation for the observed phenomenon.

- ❖ In everyday life, a hypothesis could be as simple as "drinking more water will improve my overall energy levels."

3. Prediction:

- ❖ A hypothesis leads to the formulation of testable predictions, which are specific statements that can be confirmed or refuted through experimentation.

- ❖ For example, if the hypothesis is that drinking more water increases energy levels, a prediction could be that drinking eight glasses of water per day will result in increased alertness and productivity.

4. Experimentation:

- ❖ Experiments are designed to test the predictions derived from the hypothesis.

❖ In everyday life, experimentation might involve implementing a specific action or intervention and observing its effects. For instance, one could increase their water intake for a week and track changes in energy levels.

5. Data Collection & Analysis:

❖ Data is collected through careful observation and measurement during the experiment.

❖ In daily life, data collection could involve keeping a journal or using apps to track relevant metrics such as energy levels, mood, and hydration.

6. Drawing Conclusions:

❖ Based on the results of the experiment, conclusions are drawn regarding the validity of the hypothesis.

❖ If the data supports the hypothesis, it can be accepted with confidence. If not, the hypothesis may need to be revised or discarded.

7. Iteration & Refinement:

- ❖ Science is an iterative process, and hypotheses are continually refined based on new evidence and insights.

- ❖ In everyday life, this means adjusting behaviors or decisions based on the results of previous experiments and observations.

By applying the scientific method to everyday situations, individuals can make more informed decisions, solve problems effectively, and gain a deeper understanding of the world around them. Whether it's improving personal health habits, optimizing productivity, or solving everyday challenges, the scientific method provides a reliable framework for achieving desired outcomes.

sunwave4u@gmail.com

Hypotheses, Experiments, Data Analysis & Conclusions

Hypotheses, experiments, data analysis, and conclusions are fundamental components of the scientific method, a systematic approach to inquiry used to gain knowledge and understanding about the natural world. These elements are interconnected and essential for conducting scientific research and drawing meaningful conclusions. Here's an overview of each component:

1. Hypotheses:

- A hypothesis is a testable explanation or prediction for a phenomenon or observation.

- Hypotheses are formulated based on existing knowledge, observations, and theoretical frameworks.

- They are typically stated as clear, specific statements that can be confirmed or refuted through experimentation.

- ❖ A hypothesis guides the design and conduct of experiments to test its validity.

2. Experiments:

- ❖ Experiments are controlled procedures designed to test hypotheses and gather empirical evidence.

- ❖ They involve manipulating variables (independent variables) to observe their effect on other variables (dependent variables) while keeping other factors constant.

- ❖ Experimental design is crucial for ensuring the validity and reliability of results.

- ❖ Experiments may be conducted in laboratory settings or real-life contexts, depending on the research question and objectives.

3. Data Analysis:

- ❖ Data analysis involves the systematic examination and interpretation of collected data to extract meaningful insights and draw conclusions.

- Statistical methods and techniques are commonly used to analyze data, identify patterns, trends, and relationships, and assess the significance of results.

- Data analysis may involve descriptive statistics (e.g., mean, median, standard deviation) and inferential statistics (e.g., hypothesis testing, regression analysis) to draw conclusions from the data.

4. Conclusions:

- Conclusions are drawn based on the results of data analysis and interpretation.

- They involve determining whether the evidence supports or refutes the hypothesis and what implications the findings have for the research question or problem.

- Conclusions should be based on sound reasoning, supported by empirical evidence, and acknowledge any limitations or uncertainties in the study.

- They may lead to further research questions, theoretical

developments, or practical applications in the relevant field.

Overall, hypotheses, experiments, data analysis, and conclusions are integral components of the scientific process, guiding researchers in systematically investigating phenomena, testing theories, and advancing knowledge. By understanding and applying these elements effectively, scientists can generate reliable findings and contribute to the collective understanding of the world. Additionally, individuals can apply the principles of the scientific method in various contexts to make informed decisions, solve problems, and critically evaluate information in everyday life.

1.2: Critical Thinking

Developing Critical Thinking Skills To Evaluate Information & Make Informed Decisions

Critical thinking is a foundational skill that enables individuals to assess, analyze, and evaluate information effectively to make informed decisions. In today's information-rich and complex world, honing critical thinking skills is essential for navigating diverse sources of information, identifying biases, and distinguishing between fact and opinion. Here are some key strategies to develop critical thinking skills:

1. Question Assumptions:

- ❖ Challenge assumptions underlying information or arguments by asking probing questions.

- ❖ Consider alternative perspectives and viewpoints to gain a more comprehensive understanding of the issue at hand.

2. Evaluate Sources:

- Assess the credibility, reliability, and authority of sources by examining their credentials, expertise, and evidence-based support.

- Verify information from multiple reputable sources to ensure accuracy and minimize the impact of biases or misinformation.

3. Analyze Arguments:

- Break down complex arguments into their component parts, including premises, evidence, and conclusions.

- Evaluate the logical coherence, consistency, and validity of arguments, identifying any fallacies or weaknesses in reasoning.

4. Apply Scepticism:

- Approach information with a healthy dose of scepticism, questioning claims and assertions until sufficient evidence is provided.

- Be wary of confirmation bias, the tendency to seek out or interpret information in a way that confirms pre-existing beliefs or biases.

5. Seek Evidence:

- Demand evidence to support claims and assertions, prioritizing empirical data, research findings, and verifiable facts over anecdotal or speculative information.

- Evaluate the quality, relevance, and reliability of evidence, considering factors such as sample size, methodology, and peer review.

6. Consider Context:

- Analyze information within its broader context, including historical, cultural, and social factors that may influence interpretations and perspectives.

- Recognize the nuances and complexities of issues, avoiding oversimplification or reductionism.

sunwave4u@gmail.com

7. Engage in Reflective Thinking:

- ❖ Reflect on your own thought processes, biases, and assumptions, considering how they may influence your interpretation of information and decision-making.

- ❖ Be open to revising your views in light of new evidence or alternative viewpoints, fostering intellectual humility and flexibility.

8. Practice Problem-Solving:

- ❖ Apply critical thinking skills to real-world problems and decision-making scenarios, identifying relevant information, weighing alternative courses of action, and evaluating potential consequences.

- ❖ Seek feedback and constructive criticism to improve your analytical skills and decision-making abilities over time.

By actively cultivating critical thinking skills, individuals can enhance their ability to evaluate information, discern truth

sunwave4u@gmail.com

from falsehood, and make reasoned decisions that are grounded in evidence and sound reasoning. In a world inundated with information and misinformation, critical thinking serves as a powerful tool for navigating complexity, fostering intellectual autonomy, and promoting informed citizenship.

Recognizing Biases, Fallacies & Logical Errors

In our daily lives, we encounter a multitude of information, opinions, and arguments from various sources. However, not all information is presented objectively, and not all arguments are logically sound. To navigate this landscape effectively, it's crucial to be able to recognize biases, fallacies, and logical errors. Here's how to do it:

1. Understanding Biases:

- ❖ Biases are systematic deviations from rationality or objectivity, influencing our perception, judgment, and decision-making.

- ❖ Common biases include confirmation bias (favoring information that confirms pre-existing beliefs), availability bias (overestimating the importance of information readily available), and anchoring bias (relying too heavily on initial information).

- ❖ Recognize your own biases and be mindful of how they may influence your interpretation of information and decisions.

2. Identifying Fallacies:

Fallacies are flawed arguments or reasoning patterns that undermine the validity or persuasiveness of an argument. Examples of common fallacies include:

- ❖ Ad Hominem: Attacking the person making the argument rather than addressing the argument itself.

- ❖ Straw Man: Misrepresenting or distorting an opponent's argument to make it easier to refute.

- ❖ Appeal to Authority: Asserting that a claim is true simply because an authority figure or expert says it is.

- ❖ False Dichotomy: Presenting an argument as if there are only two options when other possibilities exist.

- ❖ Familiarize yourself with different types of fallacies to spot them in arguments and avoid being misled by faulty reasoning.

3. Recognizing Logical Errors:

Logical errors occur when arguments lack coherence, consistency, or validity. Common logical errors include:

- ❖ Circular Reasoning: Using the conclusion of an argument to support its premises, thereby assuming what one is trying to prove.

- ❖ Non Sequitur: Drawing a conclusion that does not logically follow from the premises or evidence presented.

- ❖ Hasty Generalization: Making a broad generalization based on insufficient or unrepresentative evidence.

- ❖ Post Hoc Fallacy: Incorrectly assuming that because one event follows another, the first event caused the second.

- ❖ Scrutinize arguments for logical consistency and validity, ensuring that conclusions are supported by evidence and sound reasoning.

4. Applying Critical Thinking:

- ❖ Develop critical thinking skills to evaluate information and arguments critically, questioning assumptions, seeking evidence, and analyzing logical structure.

- ❖ Be open-minded and willing to consider alternative viewpoints, even if they challenge your existing beliefs or preferences.

- ❖ Practice skepticism and demand rigorous evidence to support claims, particularly those that seem too good to be true or that align closely with your biases.

By being aware of biases, fallacies, and logical errors, individuals can become more discerning consumers of information, better equipped to evaluate arguments critically and make informed decisions based on evidence and sound reasoning. Developing these skills is essential for navigating the complexities of modern discourse and promoting intellectual integrity and rationality.

Part 2: Scientific Nutrition

2.1: The Science of Nutrition

The Fundamentals Of Nutrition & The Role Of Macronutrients & Micronutrients In The Body

Nutrition is the process by which organisms obtain and utilize nutrients essential for growth, maintenance, and overall health. These nutrients can be classified into two main categories: macronutrients and micronutrients. Understanding the roles of macronutrients and micronutrients is essential for maintaining a balanced and healthy diet. Let's delve into the fundamentals of nutrition:

1. Macronutrients:

- ❖ Macronutrients are nutrients that are required by the body in large quantities to provide energy and support various physiological functions.

- ❖ The three primary macronutrients are carbohydrates, proteins, and fats.

A. Carbohydrates:

- Carbohydrates are the body's primary source of energy, providing 4 calories per gram.

- They are found in foods such as grains, fruits, vegetables, legumes, and dairy products.

- Carbohydrates are broken down into glucose, which is used by cells for energy, and glycogen, which is stored in the liver and muscles for future energy needs.

- Complex carbohydrates, such as whole grains and vegetables, are preferred over simple carbohydrates, such as refined sugars, due to their slower digestion and steady release of energy.

B. Proteins:

- Proteins are essential for building and repairing tissues, synthesizing enzymes and hormones, and supporting immune function.

- They provide 4 calories per gram and are composed of amino acids, which are often referred to as the "building blocks" of proteins.

- Protein-rich foods include meat, poultry, fish, eggs, dairy products, legumes, nuts, and seeds.

- Consuming a variety of protein sources ensures adequate intake of essential amino acids, as different foods contain different amino acid profiles.

C. Fats:

- Fats serve as a concentrated source of energy, providing 9 calories per gram.

- They are essential for hormone synthesis, cell membrane structure, and absorption of fat-soluble vitamins (A, D, E, and K).

- Healthy fats, such as monounsaturated and polyunsaturated fats found in olive oil, avocados, nuts, and fatty fish, are preferred over saturated and trans fats found in processed foods and fried foods.

- ❖ Balancing fat intake is important for maintaining cardiovascular health and overall well-being.

2. Micronutrients:

- ❖ Micronutrients are essential vitamins and minerals required by the body in smaller quantities to support various physiological processes and biochemical reactions.

- ❖ Micronutrients play crucial roles in metabolism, immune function, bone health, and other vital functions.

A. Vitamins:

- ❖ Vitamins are organic compounds that regulate metabolism, support growth and development, and act as antioxidants to protect cells from damage.

- ❖ There are two main types of vitamins: water-soluble vitamins (e.g., vitamin C and B vitamins) and fat-soluble vitamins (e.g., vitamins A, D, E, and K).

sunwave4u@gmail.com

- ❖ Vitamins are obtained from a diverse diet that includes fruits, vegetables, whole grains, dairy products, and lean proteins.

B. Minerals:

- ❖ Minerals are inorganic elements essential for various physiological functions, including bone health, nerve transmission, fluid balance, and enzyme activity.

- ❖ Major minerals, such as calcium, magnesium, sodium, potassium, phosphorus, and chloride, are needed in larger amounts, while trace minerals, such as iron, zinc, copper, selenium, and iodine, are required in smaller amounts.

- ❖ Minerals are obtained from a balanced diet that includes a variety of whole foods, including fruits, vegetables, dairy products, legumes, nuts, seeds, and lean meats.

In summary, a balanced diet that includes adequate amounts of macronutrients (carbohydrates, proteins, and fats) and micronutrients (vitamins and minerals) is essential for supporting overall health and well-being. By understanding the roles of these nutrients and making informed dietary choices, individuals can optimize their nutritional intake and promote longevity and vitality.

Debunking Common Myths & Misconceptions About Diet & Nutrition

In the realm of diet and nutrition, misinformation and misconceptions abound, often leading to confusion and misguided dietary choices. Debunking common myths is essential for promoting accurate information and helping individuals make informed decisions about their health and nutrition. Here are some prevalent myths and misconceptions about diet and nutrition, along with explanations to debunk them:

1. Myth: "Eating fat makes you fat."

Debunking: While it may seem counterintuitive, consuming dietary fat does not directly lead to weight gain. In fact, healthy fats play a crucial role in satiety, hormone regulation, and nutrient absorption. The key is to focus on consuming unsaturated fats (found in foods like avocados, nuts, seeds, and fatty fish) in moderation, while limiting intake of saturated and trans fats found in processed and fried foods.

2. Myth: "Skipping meals or fasting is the best way to lose weight."

Debunking: While intermittent fasting and occasional meal skipping may have some benefits for certain individuals, they are not universally effective weight loss strategies. Skipping meals can lead to overeating later in the day and may result in nutrient deficiencies if not properly managed. Sustainable weight loss is best achieved through a balanced diet that includes regular meals and snacks, along with physical activity.

3. Myth: "All carbs are bad for you."

Debunking: Carbohydrates are a primary source of energy for the body and play a crucial role in overall health. While refined carbohydrates (such as white bread, sugary snacks, and pastries) should be limited due to their low nutritional value and potential negative impact on blood sugar levels, complex carbohydrates (found in whole grains, fruits, vegetables, and legumes) are an important part of a healthy diet. Focus on consuming fiber-rich, nutrient-dense carbohydrates to support energy levels and overall well-being.

4. Myth: "All dietary supplements are safe & effective."

Debunking: While certain dietary supplements may provide benefits for specific populations or health conditions, not all supplements are created equal. Many supplements on the market lack scientific evidence to support their efficacy and safety. Additionally, taking high doses of certain vitamins and minerals can have adverse effects and may interact with medications. It's essential to consult with a healthcare professional before starting any new supplement regimen and to prioritize obtaining nutrients from whole foods whenever possible.

5. Myth: "You need to detox or cleanse your body regularly."

Debunking: The idea of "detoxing" or "cleansing" the body through restrictive diets, juice cleanses, or detox supplements is largely unsupported by scientific evidence. The body has its own built-in detoxification mechanisms, primarily carried out by the liver, kidneys, and lymphatic system. Instead of extreme detox programs, focus on adopting a balanced diet rich in fruits, vegetables, whole grains, lean proteins, and healthy fats, along with staying hydrated and engaging in regular physical activity, to support the body's natural detoxification processes.

By debunking these common myths and misconceptions about diet and nutrition, individuals can make more informed dietary choices and prioritize evidence-based practices for promoting health and well-being. Staying informed and critical of nutrition information can help separate fact from fiction and empower individuals to take control of their dietary habits in a meaningful and sustainable way.

sunwave4u@gmail.com

2.2: Building a Healthy Diet

Practical Tips For Designing A Balanced & Nutritious Diet Based On Scientific Evidence

Designing a balanced and nutritious diet is essential for promoting overall health and well-being. By incorporating evidence-based guidelines and principles, individuals can create a dietary pattern that provides essential nutrients, supports energy levels, and reduces the risk of chronic diseases. Here are some practical tips for designing a balanced and nutritious diet based on scientific evidence:

1. Prioritize Whole Foods:

- ❖ Base your diet on whole, minimally processed foods such as fruits, vegetables, whole grains, lean proteins, and healthy fats.

- ❖ Choose a variety of colorful fruits and vegetables to ensure a wide range of vitamins, minerals, and antioxidants.

2. Include a Variety of Nutrient-Dense Foods:

- ❖ Incorporate nutrient-dense foods that provide essential vitamins, minerals, and other beneficial compounds.

- ❖ Examples include leafy greens, berries, nuts, seeds, legumes, fatty fish, lean meats, whole grains, and dairy or plant-based alternatives.

3. Balance Macronutrients:

- ❖ Aim to include a balance of carbohydrates, proteins, and fats in your meals to support energy levels, muscle repair, and overall health.

- ❖ Choose complex carbohydrates such as whole grains, vegetables, and legumes, lean sources of protein such as poultry, fish, tofu, and beans, and healthy fats such as avocados, nuts, seeds, and olive oil.

4. Pay Attention to Portion Sizes:

- ❖ Be mindful of portion sizes to prevent overeating and maintain a healthy weight.

sunwave4u@gmail.com

- Use visual cues such as the size of your palm or a deck of cards to estimate appropriate portion sizes for proteins, grains, and fats.

5. Limit Added Sugars & Processed Foods:

- Minimize consumption of foods and beverages high in added sugars, refined grains, and unhealthy fats.

- Read food labels and ingredient lists to identify hidden sources of added sugars and artificial additives.

6. Stay Hydrated:

- Drink plenty of water throughout the day to stay hydrated and support proper bodily functions.

- Limit intake of sugary beverages, sodas, and excessive caffeine, opting for water, herbal teas, or sparkling water instead.

7. Plan & Prepare Meals:

- Plan your meals and snacks in advance to ensure balanced nutrition and avoid relying on convenience foods.

sunwave4u@gmail.com

- Prepare meals at home using fresh ingredients whenever possible, and experiment with new recipes and cooking methods to keep meals interesting and enjoyable.

8. Listen to Your Body:

- Pay attention to hunger and fullness cues, eating when hungry and stopping when satisfied.

- Practice mindful eating by savoring each bite, chewing slowly, and avoiding distractions such as screens or electronic devices during meals.

9. Seek Professional Guidance if Needed:

- Consult with a registered dietitian or nutritionist for personalized dietary recommendations tailored to your individual needs, preferences, and health goals.

- Consider any specific dietary restrictions, food allergies, or medical conditions that may require specialized guidance.

By incorporating these practical tips into your daily routine, you can design a balanced and nutritious diet that supports optimal health, energy levels, and overall well-being. Remember to focus on making gradual, sustainable changes and prioritize consistency and moderation in your dietary habits for long-term success.

sunwave4u@gmail.com

Understanding Portion Control, Meal Planning & Mindful Eating

Portion control, meal planning, and mindful eating are essential practices for fostering a healthy relationship with food, promoting balanced nutrition, and supporting overall well-being. By incorporating these strategies into your daily routine, you can optimize your dietary habits and achieve your health goals. Let's explore each of these concepts in more detail:

1. Portion Control:

- ❖ Portion control involves managing the amount of food you eat to ensure you consume appropriate serving sizes and avoid overeating.

- ❖ Use visual cues or measuring tools to estimate portion sizes, such as using your hand or common household items as references.

- ❖ Pay attention to recommended serving sizes listed on food packaging and nutrition labels, and be mindful of portion distortion at restaurants or when eating out.

- ❖ Practice portion control by serving meals on smaller plates, dividing large portions into smaller servings, and avoiding mindless eating in front of the television or computer.

2. Meal Planning:

- ❖ Meal planning involves preparing and organizing meals and snacks in advance to ensure a balanced and nutritious diet throughout the week.

- ❖ Set aside time each week to plan your meals, taking into account your schedule, dietary preferences, and nutritional goals.

- ❖ Create a grocery list based on your meal plan to streamline shopping trips and avoid impulse purchases.

- ❖ Batch cook and prepare meals in advance to save time during busy weekdays, and store leftovers for future meals or snacks.

- ❖ Experiment with meal prep containers, cooking techniques, and flavor combinations to keep meals exciting and enjoyable.

3. Mindful Eating:

- ❖ Mindful eating is the practice of being fully present and attentive to the eating experience, including the taste, texture, aroma, and sensations of food.

- ❖ Slow down and savor each bite, chewing thoroughly and paying attention to the flavors and textures of your food.

- ❖ Eat without distractions, such as watching TV or scrolling on your phone, to focus on the act of eating and tune into your body's hunger and fullness cues.

- ❖ Tune into your body's hunger and fullness signals, eating when you're hungry and stopping when you're satisfied, rather than relying on external cues or emotions.

- ❖ Cultivate a nonjudgmental attitude towards food and

yourself, allowing for flexibility and enjoyment in your eating habits without guilt or shame.

By incorporating portion control, meal planning, and mindful eating into your lifestyle, you can develop healthier eating habits, improve your relationship with food, and support your overall health and well-being. These practices empower you to make informed dietary choices, optimize nutrient intake, and cultivate a balanced and sustainable approach to nutrition. Remember that consistency and mindfulness are key, and small changes over time can lead to significant improvements in your dietary habits and overall quality of life.

Part 3: Fitness & Exercise

3.1: Exercise Physiology

Understanding How Exercise Affects The Body At The Cellular & Molecular Levels

Exercise is known to have numerous beneficial effects on the body, including improving cardiovascular health, enhancing muscular strength and endurance, and promoting overall well-being. These effects are not only evident at the physiological level but also at the cellular and molecular levels. Let's delve into how exercise influences cellular and molecular processes within the body:

1. Cellular Adaptations:

- ❖ Exercise triggers a variety of cellular adaptations in different tissues and organs throughout the body.

- ❖ In skeletal muscle tissue, exercise stimulates the production of mitochondria, the powerhouse of the cell responsible for generating energy in the form of ATP (adenosine triphosphate). This increase in

sunwave4u@gmail.com

- ❖ mitochondrial density enhances cellular energy production and improves overall metabolic efficiency.

- ❖ Exercise also promotes the growth and repair of muscle fibers through a process called hypertrophy. Regular physical activity stimulates the synthesis of muscle proteins, leading to an increase in muscle mass and strength over time.

2. Molecular Signaling Pathways:

- ❖ Exercise activates various molecular signaling pathways within cells, leading to changes in gene expression and protein synthesis.

- ❖ One key pathway activated by exercise is the AMP-activated protein kinase (AMPK) pathway, which serves as a cellular energy sensor. AMPK activation stimulates metabolic processes that enhance glucose uptake, fatty acid oxidation, and mitochondrial biogenesis, contributing to improved energy metabolism and endurance.

- ❖ Exercise also activates the mammalian target of

rapamycin (mTOR) pathway, which regulates protein synthesis and cell growth. mTOR activation promotes muscle protein synthesis and hypertrophy in response to resistance training, contributing to muscle adaptation and strength gains.

3. Hormonal Responses:

- ❖ Exercise induces changes in hormonal levels and signaling, influencing various physiological processes throughout the body.

- ❖ During exercise, the release of hormones such as adrenaline (epinephrine) and cortisol helps mobilize energy stores and increase heart rate and blood flow to support physical activity.

- ❖ Exercise also stimulates the production of endorphins, neurotransmitters that act as natural painkillers and mood enhancers, leading to feelings of euphoria and well-being commonly referred to as the "runner's high."

4. Oxidative Stress & Antioxidant Defense:

- ❖ Exercise generates reactive oxygen species (ROS) as a byproduct of increased energy metabolism and oxygen consumption.

- ❖ While ROS can have detrimental effects on cells and tissues if present in excess, regular exercise stimulates the production of endogenous antioxidants, such as superoxide dismutase (SOD) and glutathione, which help neutralize ROS and protect against oxidative damage.

- ❖ Exercise-induced oxidative stress also triggers adaptive responses that enhance antioxidant defense mechanisms and improve cellular resilience to oxidative stress over time.

In summary, exercise exerts profound effects on the body at the cellular and molecular levels, influencing cellular metabolism, gene expression, protein synthesis, hormonal responses, and oxidative stress. Understanding these mechanisms can provide insights into the physiological adaptations that occur in response to exercise and underscore the importance of regular physical activity for promoting overall health and well-being. Incorporating exercise into your daily routine can yield a wide range of benefits, from improving energy levels and metabolic function to enhancing physical performance and longevity.

Exploring Different Types Of Exercise & Their Benefits

Exercise encompasses a wide range of physical activities that vary in intensity, duration, and focus. Each type of exercise offers unique benefits for overall health, fitness, and well-being. By incorporating a variety of exercises into your routine, you can target different muscle groups, improve cardiovascular health, enhance flexibility, and promote mental and emotional well-being. Here are some common types of exercise and their respective benefits:

1. Aerobic Exercise:

Aerobic exercise, also known as cardio or cardiovascular exercise, involves activities that increase your heart rate and breathing rate for an extended period.

Benefits:

- ❖ Improves cardiovascular health by strengthening the heart muscle and enhancing circulation.

- ❖ Increases lung capacity and improves respiratory function.

- ❖ Helps manage weight by burning calories and promoting fat loss.

- ❖ Reduces the risk of chronic diseases such as heart disease, diabetes, and stroke.

- ❖ Boosts mood and reduces stress through the release of endorphins.

- ❖ Examples: Running, walking, cycling, swimming, dancing, aerobic classes, and HIIT (high-intensity interval training).

2. Strength Training:

Strength training involves resistance exercises designed to build muscle strength, endurance, and power.

Benefits:

- ❖ Increases muscle mass and improves muscular strength and endurance.

- ❖ Boosts metabolism, leading to greater calorie expenditure and fat loss.

- ❖ Enhances bone density and reduces the risk of osteoporosis.
- ❖ Improves functional strength for daily activities and sports performance.

- ❖ Supports joint health and reduces the risk of injury.

- ❖ Examples: Weightlifting, bodyweight exercises (e.g., push-ups, squats, lunges), resistance band workouts, and kettlebell training.

3. Flexibility & Stretching:

Flexibility exercises focus on improving the range of motion and elasticity of muscles and joints.

Benefits:

- ❖ Increases flexibility and mobility, allowing for greater freedom of movement.

- Reduces muscle tension and stiffness, alleviating discomfort and improving posture.

- Helps prevent injuries by improving joint stability and reducing the risk of strains and sprains.

- Promotes relaxation and reduces stress by releasing tension in the body.

- Examples: Yoga, Pilates, static stretching, dynamic stretching, and mobility drills.

4. Balance & Stability Training:

Balance and stability exercises help improve proprioception, coordination, and postural control.

Benefits:

- Enhances balance and coordination, reducing the risk of falls and injuries, especially in older adults.

- Strengthens stabilizing muscles and improves core strength, leading to better posture and spinal alignment.

- ❖ Improves athletic performance by enhancing proprioceptive awareness and agility.

- ❖ Enhances body awareness and mindfulness.

- ❖ Examples: Tai Chi, Qigong, balance exercises (e.g., standing on one leg, stability ball exercises), and proprioceptive training drills.

5. Mind-Body Practices:

Mind-body practices combine physical movement with mental focus and relaxation techniques to promote holistic well-being.

Benefits:

- ❖ Reduces stress, anxiety, and depression by promoting relaxation and mindfulness.

- ❖ Improves mental clarity, focus, and cognitive function.

- ❖ Enhances self-awareness and emotional regulation.

- ❖ Strengthens the mind-body connection and promotes overall resilience.

- ❖ Examples: Yoga, Tai Chi, Qigong, Pilates, and mindful movement practices.

Incorporating a variety of exercises into your routine can help you achieve a well-rounded fitness regimen that targets different aspects of physical and mental health. Aim for a balanced approach that includes aerobic exercise, strength training, flexibility work, balance and stability training, and mind-body practices to maximize the benefits of exercise and support overall well-being. Remember to listen to your body, start gradually, and consult with a healthcare professional before beginning any new exercise program, especially if you have underlying health concerns or medical conditions.

3.2: Designing an Effective Workout Routine

Guidelines For Creating A Personalized Workout Routine That Aligns With Your Fitness Goals

Creating a personalized workout routine involves tailoring your exercise program to align with your specific fitness goals, preferences, and individual needs. Whether you're aiming to build strength, improve cardiovascular fitness, lose weight, or enhance overall well-being, designing a workout plan that suits your objectives is essential for success. Here are some guidelines to help you create a personalized workout routine:

1. Define Your Fitness Goals:

- ❖ Start by clearly defining your fitness goals. Are you looking to build muscle, increase endurance, lose weight, improve flexibility, or enhance overall fitness and well-being?

- ❖ Set specific, measurable, achievable, relevant, and

time-bound (SMART) goals to guide your exercise program and track progress over time.

2. Assess Your Current Fitness Level:

- ❖ Evaluate your current fitness level, including strength, cardiovascular endurance, flexibility, and mobility.

- ❖ Consider any limitations, injuries, or medical conditions that may impact your ability to exercise and adapt your routine accordingly.

3. Choose the Right Types of Exercise:

- ❖ Select exercises and activities that align with your fitness goals and preferences.

- ❖ Include a variety of aerobic, strength training, flexibility, balance, and functional movement exercises to achieve a well-rounded workout routine.

- ❖ Incorporate activities that you enjoy and look forward to, as this will increase adherence and motivation.

4. Determine Frequency, Intensity & Duration:

- ❖ Decide how often you will exercise each week based on your schedule, fitness goals, and recovery needs.

- ❖ Determine the intensity of your workouts, considering factors such as heart rate, perceived exertion, and resistance level.

- ❖ Establish the duration of each workout session, balancing effectiveness with time constraints and recovery considerations.

5. Plan Your Workouts:

- ❖ Structure your workout routine by organizing exercises into a weekly schedule or split routine.

- ❖ Include a warm-up and cool-down period in each workout to prepare the body for exercise and aid in recovery.

- Rotate between different types of exercises and muscle groups to prevent boredom, overuse injuries, and plateaus.

- Gradually progress the intensity, volume, and complexity of your workouts over time to challenge your body and stimulate adaptation.

6. Listen to Your Body:

- Pay attention to how your body responds to exercise and adjust your routine accordingly.

- Monitor signs of fatigue, soreness, and overtraining, and incorporate rest days, active recovery, and deload periods as needed.

- Be mindful of any pain or discomfort during exercise and modify or avoid activities that exacerbate existing injuries or conditions.

7. Track Your Progress:

- Keep a workout log or journal to record your exercises, sets, reps, weights, and other relevant details.

sunwave4u@gmail.com

- ❖ Monitor changes in fitness metrics, such as strength gains, improvements in cardiovascular endurance, changes in body composition, and increases in flexibility and mobility.

- ❖ Celebrate achievements and milestones along the way to stay motivated and committed to your fitness journey.

8. Seek Professional Guidance:

- ❖ Consider working with a certified personal trainer, fitness coach, or exercise specialist to develop a customized workout plan tailored to your goals and needs.

- ❖ Consult with a healthcare professional, especially if you have underlying health concerns or medical conditions, to ensure that your exercise program is safe and appropriate.

By following these guidelines and creating a personalized workout routine that aligns with your fitness goals, preferences, and individual needs, you can maximize the effectiveness of your exercise program and achieve lasting results. Remember that consistency, dedication, and patience are key to success, and embrace the journey towards a healthier, stronger, and happier you.

Incorporating Strength Training, Cardiovascular Exercise, Flexibility & Recovery

A well-rounded workout routine should encompass a combination of strength training, cardiovascular exercise, flexibility work, and recovery strategies to promote overall fitness, health, and well-being. By incorporating each of these components into your routine, you can achieve a balanced and sustainable approach to exercise that supports your fitness goals and enhances your quality of life. Here's how to integrate strength training, cardiovascular exercise, flexibility work, and recovery into your workout routine:

1. Strength Training:

- ❖ Aim to perform strength training exercises at least 2-3 times per week, targeting all major muscle groups.

- ❖ Include a variety of compound exercises (e.g., squats, deadlifts, bench presses) that engage multiple muscle groups simultaneously.

- ❖ Choose a weight or resistance level that allows you to perform 8-12 repetitions with proper form, aiming for 2-3 sets of each exercise.

- ❖ Progressively overload your muscles by gradually increasing the weight, reps, or sets over time to stimulate muscle growth and strength gains.

- ❖ Allow for adequate rest between strength training sessions to allow muscles to recover and adapt.

2. Cardiovascular Exercise:

- ❖ Incorporate cardiovascular exercise into your routine 3-5 times per week, aiming for at least 150 minutes of moderate-intensity aerobic activity or 75 minutes of vigorous-intensity aerobic activity per week.

- ❖ Choose activities that elevate your heart rate and increase breathing rate, such as brisk walking, running, cycling, swimming, or dancing.

- ❖ Vary the intensity, duration, and type of cardiovascular exercise to prevent boredom and maximize benefits.

- ❖ Incorporate interval training or high-intensity interval training (HIIT) to boost calorie burn, improve cardiovascular fitness, and enhance metabolic efficiency.

- ❖ Allow for adequate recovery between intense cardio sessions to avoid overtraining and reduce the risk of injury.

3. Flexibility Work:

- ❖ Dedicate time to flexibility exercises and stretching at least 2-3 times per week to improve range of motion, mobility, and muscle flexibility.

- ❖ Include dynamic stretches as part of your warm-up routine to prepare the body for exercise and static stretches at the end of your workout to cool down and promote relaxation.

- ❖ Focus on stretching all major muscle groups, paying particular attention to areas that tend to be tight or prone to stiffness.

sunwave4u@gmail.com

- Hold each stretch for 15-30 seconds and aim to feel a gentle stretch without discomfort or pain.

- Incorporate activities such as yoga, Pilates, or mobility drills to improve flexibility, balance, and body awareness.

4. Recovery Strategies:

- Prioritize rest and recovery as an essential part of your workout routine to allow your body to repair and rebuild muscle tissue, replenish energy stores, and reduce the risk of overuse injuries.

- Ensure adequate sleep of 7-9 hours per night to support physical and mental recovery, hormone regulation, and overall health.

- Incorporate active recovery days into your routine with low-intensity activities such as walking, swimming, or gentle yoga to promote circulation and muscle relaxation.

- ❖ Use foam rolling, massage, or other self-myofascial release techniques to alleviate muscle soreness and tension.

- ❖ Practice stress management techniques such as meditation, deep breathing, or mindfulness to reduce mental and emotional stress and promote relaxation.

By incorporating strength training, cardiovascular exercise, flexibility work, and recovery strategies into your workout routine, you can achieve a well-rounded approach to fitness that optimizes health, performance, and longevity. Remember to listen to your body, prioritize consistency and balance, and adjust your routine as needed to support your goals and overall well-being.

Part 4: Mental Health & Well-being

4.1: The Science of Happiness

Exploring The Psychology Of Happiness & The Factors That Contribute To Well-Being

Happiness is a complex and multifaceted emotion that encompasses feelings of pleasure, contentment, and satisfaction with life. While happiness is influenced by genetic, environmental, and situational factors, research in positive psychology has identified several key determinants and contributing factors that play a significant role in promoting well-being. Let's explore the psychology of happiness and some of the factors that contribute to overall well-being:

1. Positive Emotions:

- ❖ Positive emotions, such as joy, gratitude, love, and serenity, are fundamental components of happiness and well-being.

- ❖ Cultivating positive emotions through activities such as

- ❖ practicing gratitude, savoring pleasant experiences, and engaging in acts of kindness can enhance overall happiness and life satisfaction.

2. Engagement & Flow:

- ❖ Engagement refers to the state of being fully absorbed and immersed in an activity, often characterized by focused attention, intrinsic motivation, and a sense of enjoyment.

- ❖ Experiencing flow, a concept introduced by psychologist Mihaly Csikszentmihalyi, occurs when individuals are completely absorbed in a challenging task that matches their skills, leading to a state of optimal performance and satisfaction.

- ❖ Engaging in activities that foster flow, such as pursuing hobbies, participating in creative endeavors, or pursuing meaningful work, can contribute to feelings of fulfillment and well-being.

3. Meaning & Purpose:

- ❖ Finding meaning and purpose in life involves identifying personal values, goals, and aspirations that provide a sense of direction and significance.

- ❖ Living a purpose-driven life involves aligning one's actions and decisions with core values and contributing to something greater than oneself, whether through meaningful work, relationships, or community involvement.

- ❖ Having a sense of purpose has been linked to greater resilience, well-being, and overall life satisfaction.

4. Positive Relationships:

- ❖ Social connections and supportive relationships are crucial for happiness and well-being.

- ❖ Maintaining close, meaningful relationships with family, friends, and community members provides emotional support, validation, and a sense of belonging.

- Investing time and effort into nurturing and strengthening relationships, practicing empathy and active listening, and fostering trust and reciprocity can enhance social connectedness and happiness.

5. Accomplishment & Achievement:

- Setting and achieving goals, both short-term and long-term, can contribute to feelings of accomplishment, self-efficacy, and happiness.

- Pursuing goals that are personally meaningful and challenging, while also attainable, provides a sense of purpose and direction in life.

- Celebrating progress and milestones along the way, regardless of their size, reinforces feelings of competence and satisfaction.

6. Resilience & Coping Skills:

- Resilience refers to the ability to adapt and bounce back from adversity, setbacks, and challenges.

- Cultivating resilience involves developing coping skills, emotional regulation strategies, and positive thinking patterns that promote psychological well-being and adaptive responses to stress.

- Building resilience through practices such as mindfulness meditation, cognitive reframing, and seeking social support can enhance overall happiness and psychological health.

7. Self-Care & Well-Being Practices:

- Prioritizing self-care and well-being practices is essential for maintaining happiness and resilience in the face of life's stressors.

- Engaging in activities that promote physical health, such as regular exercise, nutritious eating, adequate sleep, and relaxation techniques, supports overall well-being and vitality.

- Practicing self-compassion, self-acceptance, and self-care fosters a positive relationship with oneself and promotes psychological resilience and happiness.

sunwave4u@gmail.com

In summary, the psychology of happiness is influenced by a variety of factors, including positive emotions, engagement, meaning and purpose, positive relationships, accomplishment and achievement, resilience and coping skills, and self-care practices. By cultivating these factors in our lives and prioritizing activities that promote well-being, individuals can enhance their happiness, satisfaction, and overall quality of life. Remember that happiness is a journey, not a destination, and that small changes and intentional choices can have a significant impact on our well-being over time.

Strategies For Cultivating Positive Emotions & Resilience.

Cultivating positive emotions and resilience is essential for promoting well-being, managing stress, and navigating life's challenges with greater ease and adaptability. By incorporating strategies to foster positive emotions and build resilience into your daily life, you can enhance your overall happiness and psychological health. Here are some effective strategies for cultivating positive emotions and resilience:

1. Practice Gratitude:

- ❖ Regularly express gratitude for the blessings and positive aspects of your life, whether big or small.

- ❖ Keep a gratitude journal and write down three things you are grateful for each day.

- ❖ Cultivate an attitude of appreciation and focus on the positives, even during difficult times.

2. Engage in Acts of Kindness:

- ❖ Perform acts of kindness towards others, such as volunteering, helping a friend in need, or expressing kindness and compassion in your interactions.

- ❖ Random acts of kindness not only benefit others but also boost your own mood and sense of well-being.

3. Foster Positive Relationships:

- ❖ Nurture supportive and meaningful relationships with family, friends, and community members.

- ❖ Invest time and effort into building and maintaining connections with others, practicing empathy, active listening, and communication skills.

- ❖ Seek out social support during times of stress or adversity, and offer support to others in return.

sunwave4u@gmail.com

4. Practice Mindfulness & Meditation:

- ❖ Incorporate mindfulness practices into your daily routine, such as meditation, deep breathing exercises, or mindful movement (e.g., yoga, Tai Chi).

- ❖ Cultivate present-moment awareness and nonjudgmental acceptance of your thoughts, feelings, and experiences.

- ❖ Mindfulness practices help reduce stress, enhance emotional regulation, and promote resilience in the face of challenges.

5. Develop Optimism & Positive Thinking:

- ❖ Foster an optimistic outlook by reframing negative thoughts and focusing on hopeful and optimistic perspectives.

- ❖ Challenge negative self-talk and cognitive distortions, replacing them with more balanced and constructive thoughts.

❖ Cultivate a growth mindset, viewing setbacks and failures as opportunities for learning and growth.

6. Cultivate Self-Compassion:

❖ Practice self-compassion by treating yourself with kindness, understanding, and acceptance, especially during difficult times.

❖ Offer yourself the same compassion and support that you would extend to a friend facing similar challenges.

❖ Practice self-care and prioritize activities that promote physical, emotional, and mental well-being.

7. Build Resilience Skills:

❖ Develop coping skills and resilience strategies to navigate adversity and bounce back from setbacks.

❖ Identify and utilize your strengths and resources to cope with stress and challenges effectively.

❖ Foster a sense of purpose and meaning in life, drawing

on personal values and goals to guide your actions and decisions.

8. Seek Support When Needed:

- ❖ Reach out for support from friends, family, or mental health professionals when facing difficult emotions or situations.

- ❖ Asking for help is a sign of strength, not weakness, and can provide valuable support and resources during challenging times.

By incorporating these strategies into your daily life, you can cultivate positive emotions, enhance resilience, and foster greater well-being and happiness. Remember that building these skills takes time and practice, so be patient and kind to yourself as you embark on your journey toward greater positivity and resilience.

4.2: Stress Management

Understanding The Physiological & Psychological Effects Of Stress On The Body

Stress is a natural response to challenging or threatening situations, triggering a complex cascade of physiological and psychological changes designed to help us cope with perceived threats. While acute stress can be adaptive and mobilize resources to deal with immediate challenges, chronic or prolonged stress can have detrimental effects on both physical and mental health. Let's explore the physiological and psychological effects of stress on the body:

1. Physiological Effects of Stress:

- ❖ Activation of the Sympathetic Nervous System: When faced with a stressor, the body's sympathetic nervous system is activated, leading to the release of stress hormones such as adrenaline (epinephrine) and cortisol.

- ❖ Increased Heart Rate and Blood Pressure: Stress

- hormones cause the heart to beat faster and blood vessels to constrict, increasing blood pressure and redirecting blood flow to vital organs.

- Enhanced Alertness and Energy: Stress hormones prepare the body for action by increasing alertness, arousal, and energy levels, enabling quick responses to perceived threats.

- Suppression of Non-Essential Functions: During periods of stress, the body prioritizes immediate survival needs, temporarily suppressing non-essential functions such as digestion, immune function, and reproductive processes.

- Activation of the HPA Axis: Chronic stress activates the hypothalamic-pituitary-adrenal (HPA) axis, leading to prolonged release of cortisol, which can have wide-ranging effects on metabolism, immune function, and inflammation.

2. Psychological Effects of Stress:

- Emotional Distress: Stress can trigger a range of

emotional responses, including anxiety, irritability, frustration, anger, and sadness.

- ❖ Cognitive Impairment: Prolonged stress can impair cognitive function, attention, memory, and decision-making, making it difficult to focus, concentrate, and perform tasks effectively.

- ❖ Negative Thinking Patterns: Chronic stress can fuel negative thinking patterns, such as catastrophic thinking, pessimism, and rumination, leading to increased feelings of helplessness and hopelessness.

- ❖ Sleep Disturbances: Stress can disrupt sleep patterns, leading to difficulty falling asleep, staying asleep, or experiencing restful sleep. Poor sleep further exacerbates stress and impairs cognitive function and mood regulation.

- ❖ Behavioral Changes: Stress can influence behavior, leading to changes such as increased irritability, social withdrawal, overeating, undereating, substance abuse, or engaging in other unhealthy coping mechanisms.

3. Long-Term Health Consequences:

- ❖ Chronic stress has been linked to a variety of long-term health consequences, including cardiovascular disease, hypertension, diabetes, obesity, gastrointestinal disorders, immune dysfunction, and mental health disorders such as depression and anxiety.

- ❖ Prolonged activation of the stress response can contribute to chronic inflammation, oxidative stress, and dysregulation of various physiological systems, increasing the risk of chronic diseases and premature aging.

- ❖ Chronic stress also affects brain structure and function, particularly regions involved in emotion regulation, memory, and stress response, potentially increasing susceptibility to mental health disorders and cognitive decline.

In summary, stress exerts profound physiological and psychological effects on the body, influencing a wide range of physiological processes, cognitive functions, emotional responses, and health outcomes. While acute stress is a normal part of life and can be adaptive in moderation, chronic or prolonged stress can have detrimental effects on both physical and mental well-being. Therefore, it's essential to develop effective stress management strategies and prioritize self-care to mitigate the negative impact of stress on overall health and quality of life.

Techniques For Managing Stress & Promoting Relaxation, Including Mindfulness & Meditation

Managing stress and promoting relaxation are essential for maintaining overall well-being, reducing the negative effects of chronic stress, and enhancing resilience in the face of life's challenges. Incorporating relaxation techniques into your daily routine can help calm the mind, reduce tension in the body, and foster a greater sense of balance and inner peace. Here are some effective techniques for managing stress and promoting relaxation, including mindfulness and meditation:

1. Mindfulness Meditation:

- ❖ Mindfulness meditation involves paying attention to the present moment with openness, curiosity, and acceptance, without judgment.

- ❖ Find a quiet and comfortable space to sit or lie down, close your eyes, and focus your attention on your breath or a specific anchor such as a word or sensation.

- ❖ Notice any thoughts, feelings, or sensations that arise without getting caught up in them, gently bringing your attention back to the present moment whenever your mind wanders.

- ❖ Practice mindfulness meditation regularly, starting with short sessions of 5-10 minutes and gradually increasing the duration as you become more comfortable with the practice.

2. Deep Breathing Exercises:

- ❖ Deep breathing exercises help activate the body's relaxation response, reducing stress and promoting a sense of calm and relaxation.

- ❖ Find a comfortable seated or lying position and place one hand on your abdomen and the other on your chest.

- ❖ Inhale deeply through your nose, allowing your abdomen to rise as you fill your lungs with air. Exhale slowly and completely through your mouth, feeling your abdomen fall.

- Continue deep breathing for several minutes, focusing on the sensation of your breath moving in and out of your body.

3. Progressive Muscle Relaxation (PMR):

- Progressive muscle relaxation involves systematically tensing and relaxing different muscle groups in the body to release physical tension and promote relaxation.

- Start by finding a quiet and comfortable space to lie down or sit in a relaxed position.

- Begin by tensing a specific muscle group (e.g., your hands, arms, shoulders) for 5-10 seconds, then release the tension and allow the muscles to relax completely.

- Move through each muscle group in your body, progressively working your way from head to toe, paying attention to the sensations of tension and relaxation.

4. Guided Imagery:

- ❖ Guided imagery involves visualizing peaceful and calming scenes or scenarios to evoke feelings of relaxation and well-being.

- ❖ Find a quiet and comfortable space to sit or lie down and close your eyes.

- ❖ Imagine yourself in a serene and tranquil setting, such as a beach, forest, or meadow, focusing on the sights, sounds, and sensations of your imagined surroundings.

- ❖ Engage all your senses as you immerse yourself in the visualization, allowing yourself to feel calm, centered, and at peace.

5. Body Scan Meditation:

- ❖ Body scan meditation involves systematically directing your attention to different parts of your body, observing any sensations or tension present, and allowing them to release and relax.

- ❖ Lie down in a comfortable position and close your eyes, bringing your awareness to your breath.

- ❖ Slowly scan your body from head to toe, noticing any areas of tension, discomfort, or tightness.

- ❖ As you encounter areas of tension, breathe into them, allowing them to soften and release with each exhale.

6. Yoga & Tai Chi:

- ❖ Yoga and Tai Chi are mind-body practices that combine gentle movements, breathwork, and mindfulness to promote relaxation, flexibility, and stress reduction.

- ❖ Practice yoga poses (asanas) or Tai Chi sequences regularly to improve physical and mental well-being, reduce muscle tension, and enhance relaxation.

- ❖ Choose styles of yoga or Tai Chi that emphasize gentle, restorative, or yin practices to support relaxation and stress relief.

7. Nature Walks & Outdoor Activities:

- Spending time in nature and engaging in outdoor activities can have a calming and rejuvenating effect on the mind and body.

- Take regular nature walks in parks, forests, or natural landscapes, paying attention to the sights, sounds, and sensations of the natural environment.

- Participate in outdoor activities such as hiking, gardening, or birdwatching to connect with nature, reduce stress, and promote relaxation.

8. Journaling & Expressive Writing:

- Journaling and expressive writing provide an outlet for processing emotions, thoughts, and experiences, reducing rumination and promoting self-awareness and emotional regulation.

- Set aside time each day to write freely about your thoughts, feelings, and experiences without judgment or censorship.

- Use journaling prompts or guided exercises to explore specific themes, express gratitude, or set intentions for the day.

9. Social Support & Connection:

- Cultivate supportive relationships and connections with friends, family, or community members who provide emotional support, understanding, and validation.

- Share your thoughts and feelings with trusted individuals, seek advice or perspective when needed, and offer support to others in return.

- Engage in social activities and meaningful interactions that foster a sense of belonging and connection.

10. Healthy Lifestyle Habits:

- Prioritize self-care and adopt healthy lifestyle habits that support overall well-being, including regular exercise, nutritious eating, adequate sleep, and stress management techniques.

- ❖ Maintain a balanced and nourishing diet rich in whole foods, stay hydrated, and limit consumption of caffeine, alcohol, and processed foods.

- ❖ Establish a consistent sleep routine and prioritize restful sleep, aiming for 7-9 hours of quality sleep per night.

- ❖ Engage in regular physical activity and exercise, incorporating a mix of cardiovascular, strength training, and flexibility exercises into your routine.

- ❖ Practice stress management techniques such as mindfulness, meditation, deep breathing, or progressive muscle relaxation to reduce stress and promote relaxation.

- ❖ Prioritize activities that bring you joy, fulfillment, and a sense of purpose, whether it's pursuing hobbies, spending time with loved ones, or engaging in creative endeavors.

By incorporating these techniques for managing stress and promoting relaxation into your daily routine, you can cultivate greater resilience, enhance well-being, and improve your ability to cope with life's challenges. Experiment with different strategies to find what works best for you, and remember that consistency and practice are key to reaping the benefits of relaxation and stress management techniques.

Part 5: Environmental Sustainability

5.1: Sustainable Living

The Importance Of Environmental Sustainability For Personal & Planetary Health

Environmental sustainability refers to the responsible use and management of natural resources to meet current needs without compromising the ability of future generations to meet their own needs. It encompasses practices that promote the conservation of biodiversity, protection of ecosystems, reduction of pollution, and mitigation of climate change. Environmental sustainability is critically important for both personal and planetary health for several reasons:

1. Protection of Ecosystem Services:

- ❖ Ecosystems provide essential services that support human well-being, including clean air and water, fertile soil, pollination of crops, and regulation of climate and weather patterns.

- ❖ Maintaining healthy ecosystems through sustainable

practices ensures the continued availability of these ecosystem services, which are vital for human health and survival.

2. Preservation of Biodiversity:

- ❖ Biodiversity is the variety of life on Earth, including plants, animals, and microorganisms, and the ecosystems they inhabit.

- ❖ Biodiversity supports ecosystem stability, resilience, and productivity, contributing to ecosystem services and providing genetic resources for food, medicine, and other essential products.

- ❖ Protecting biodiversity through sustainable land use, conservation efforts, and habitat restoration is crucial for maintaining ecosystem health and resilience in the face of environmental changes.

3. Mitigation of Climate Change:

- ❖ Environmental sustainability plays a key role in mitigating climate change, which poses significant risks to human health, ecosystems, and economies.

- Greenhouse gas emissions from human activities, such as burning fossil fuels and deforestation, contribute to global warming and climate disruption.

- Implementing sustainable practices to reduce carbon emissions, promote renewable energy sources, improve energy efficiency, and enhance carbon sequestration in forests and soils can help mitigate climate change and its adverse impacts on health and well-being.

4. Protection of Human Health:

- Environmental degradation and pollution pose serious risks to human health, leading to respiratory diseases, waterborne illnesses, food insecurity, and other health problems.

- Exposure to air pollution, contaminated water, hazardous chemicals, and toxic substances can increase the risk of respiratory disorders, cardiovascular diseases, cancer, and other health issues.

- Promoting environmental sustainability through clean energy, pollution prevention, waste reduction, and sustainable agriculture can improve air and water quality, reduce exposure to harmful chemicals, and protect human health.

5. Sustainable Resource Management:

- ❖ Unsustainable consumption and depletion of natural resources, such as freshwater, forests, fisheries, and minerals, threaten the availability of essential resources for future generations.

- ❖ Adopting sustainable resource management practices, such as conservation, recycling, and responsible stewardship of natural resources, ensures their long-term availability and supports sustainable development.

6. Social Equity & Economic Prosperity:

- ❖ Environmental sustainability is closely linked to social equity and economic prosperity, as environmental degradation often disproportionately affects marginalized communities and exacerbates social inequalities.

- ❖ Sustainable development promotes equitable access to resources, opportunities, and benefits for all members of society, ensuring that no one is left behind.

- ❖ Investing in green technologies, sustainable infrastructure, and renewable energy creates economic opportunities, promotes innovation, and fosters long-term prosperity while protecting the environment.

In summary, environmental sustainability is essential for both personal and planetary health, as it ensures the preservation of ecosystems, biodiversity, and essential ecosystem services, mitigates climate change, protects human health, promotes sustainable resource management, and fosters social equity and economic prosperity. By adopting sustainable practices and supporting policies that prioritize environmental protection and conservation, individuals and societies can contribute to a healthier, more resilient, and sustainable future for all.

Practical Tips For Reducing Your Carbon Footprint & Living More Sustainably

Reducing your carbon footprint and living more sustainably involves making conscious choices to minimize your environmental impact and promote conservation of natural resources. By adopting sustainable practices in various aspects of your daily life, you can contribute to mitigating climate change, protecting ecosystems, and fostering a healthier planet for future generations. Here are some practical tips for reducing your carbon footprint and living more sustainably:

1. Reduce Energy Consumption:

- ❖ Use energy-efficient appliances and lighting fixtures to reduce electricity consumption at home.

- ❖ Turn off lights, electronics, and appliances when not in use, and unplug chargers and devices to avoid standby power consumption.

- Adjust your thermostat to conserve energy, using programmable thermostats or smart thermostats to optimize heating and cooling settings.

- Use natural lighting and ventilation whenever possible, and consider installing energy-efficient windows and insulation to improve energy efficiency.

2. Conserve Water:

- Fix leaks and drips in faucets, toilets, and pipes to prevent water waste.

- Install water-saving devices such as low-flow showerheads, faucet aerators, and dual-flush toilets to reduce water usage.

- Take shorter showers, turn off the tap while brushing teeth or washing dishes, and collect rainwater for outdoor use.

- Use drought-resistant landscaping and practice water-wise gardening techniques to conserve water in landscaping and gardening.

3. Reduce, Reuse, Recycle:

- ❖ Minimize waste by reducing consumption, reusing items whenever possible, and recycling materials such as paper, cardboard, glass, plastic, and metal.

- ❖ Opt for reusable alternatives to single-use items, such as reusable shopping bags, water bottles, coffee cups, and food containers.

- ❖ Compost organic waste, such as food scraps and yard debris, to divert organic matter from landfills and produce nutrient-rich compost for gardening.

4. Choose Sustainable Transportation:

- ❖ Reduce car dependency by walking, cycling, carpooling, or using public transit for commuting and errands.

- ❖ Opt for fuel-efficient vehicles, car-sharing services, or electric vehicles (EVs) to minimize greenhouse gas emissions from transportation.

sunwave4u@gmail.com

- Combine trips, plan routes efficiently, and practice eco-driving techniques to reduce fuel consumption and vehicle emissions.

5. Eat Sustainably:

- Choose locally sourced, seasonal, and organic foods whenever possible to support local farmers and reduce carbon emissions associated with food transportation.

- Reduce meat consumption and incorporate more plant-based meals into your diet to reduce the environmental impact of food production, such as land use, water usage, and greenhouse gas emissions.

- Minimize food waste by planning meals, storing food properly, and composting food scraps.

6. Support Sustainable Products & Practices:

- Purchase products that are eco-friendly, sustainably produced, and made from recycled or renewable materials.

- ❖ Support companies and brands that prioritize environmental sustainability, ethical labor practices, and social responsibility.

- ❖ Choose products with minimal packaging or packaging that is recyclable, compostable, or biodegradable.

7. Conserve Resources:

- ❖ Practice mindful consumption and avoid purchasing unnecessary items or products with excessive packaging.

- ❖ Repair, repurpose, or donate items instead of discarding them, and consider borrowing or renting items that you only need temporarily.

- ❖ Invest in durable, high-quality products that are built to last and require fewer replacements over time.

8. Advocate for Change:

- ❖ Get involved in local and community-based initiatives that promote environmental sustainability, such as clean-up events, recycling programs, and conservation projects.

- ❖ Support policies and legislation that prioritize environmental protection, renewable energy, and sustainable development at the local, national, and global levels.

- ❖ Educate others about the importance of reducing carbon footprints and living more sustainably, and encourage collective action to address environmental challenges.

By incorporating these practical tips for reducing your carbon footprint and living more sustainably into your daily life, you can make a positive impact on the environment and contribute to building a more sustainable and resilient future for all. Remember that small changes add up, and every effort counts towards creating a healthier planet for current and future generations.

5.2: Consumer Choices

Making Environmentally Conscious Choices When Purchasing Food, Clothing, Household Products & Electronics

Making environmentally conscious choices when purchasing everyday items is an effective way to reduce your carbon footprint, minimize waste, and support sustainable practices. By considering the environmental impact of the products you buy, you can make informed decisions that contribute to conservation efforts and promote eco-friendly alternatives. Here are some tips for making environmentally conscious choices when purchasing food, clothing, household products, and electronics:

1. Food:

- ❖ Choose locally grown and seasonal produce whenever possible to reduce carbon emissions associated with transportation and support local farmers.

- Opt for organic foods to minimize exposure to pesticides and synthetic chemicals, and to support sustainable agricultural practices that protect soil and water quality.

- Select sustainably sourced seafood that is certified by reputable organizations such as the Marine Stewardship Council (MSC) or Aquaculture Stewardship Council (ASC) to ensure responsible fishing and aquaculture practices.

- Reduce meat consumption and choose plant-based alternatives to lessen the environmental impact of animal agriculture, such as deforestation, greenhouse gas emissions, and water usage.

- Look for eco-friendly packaging options, such as products with minimal or recyclable packaging, bulk bins, or reusable containers, to minimize waste and reduce plastic pollution.

2. Clothing:

- Prioritize quality over quantity and invest in durable,

timeless clothing that will last longer and require fewer replacements, reducing overall consumption and waste.

- ❖ Choose clothing made from sustainable and eco-friendly materials, such as organic cotton, bamboo, hemp, linen, or recycled fibers, which have lower environmental impacts compared to conventional fabrics.

- ❖ Support ethical and fair trade fashion brands that prioritize worker rights, fair wages, and safe working conditions throughout the supply chain.

- ❖ Consider shopping secondhand or vintage clothing to extend the lifespan of garments and reduce the demand for new clothing production, thereby reducing resource consumption and waste.

- ❖ Properly care for and maintain your clothing by washing in cold water, air-drying when possible, and repairing or repurposing items to prolong their useable life.

3. Household Products:

❖ Choose eco-friendly cleaning products that are non-toxic, biodegradable, and free from harmful chemicals such as phthalates, parabens, and chlorine bleach.

❖ Look for household products with eco-label certifications, such as the Environmental Protection Agency's Safer Choice label or third-party certifications like Green Seal or EcoLogo, which indicate products that meet environmental and health standards.

❖ Use concentrated or refillable cleaning products to reduce packaging waste and minimize the environmental impact of transporting water-heavy products.

❖ Opt for reusable and sustainable alternatives to single-use disposable items, such as cloth napkins, reusable food storage containers, and refillable soap dispensers, to reduce waste and conserve resources.

❖ Choose energy-efficient appliances and electronics with high ENERGY STAR ratings to minimize energy consumption and reduce greenhouse gas emissions associated with electricity use.

4. Electronics:

- ❖ Consider the environmental impact of electronics when making purchasing decisions, including factors such as energy efficiency, product lifespan, recyclability, and the use of hazardous materials.

- ❖ Choose electronics with high energy efficiency ratings and low standby power consumption to minimize energy use and reduce electricity costs over time.

- ❖ Look for electronics with eco-friendly certifications, such as Electronic Product Environmental Assessment Tool (EPEAT) or ENERGY STAR, which indicate products that meet environmental performance standards.

- ❖ Extend the lifespan of electronics by properly maintaining and repairing them when necessary, rather than replacing them prematurely.

❖ Recycle old electronics responsibly through certified e-waste recycling programs to recover valuable materials and prevent hazardous substances from entering the environment.

By applying these tips for making environmentally conscious choices when purchasing food, clothing, household products, and electronics, you can reduce your environmental impact, support sustainable practices, and contribute to a healthier planet for current and future generations. Remember that small changes in purchasing habits can make a big difference over time, and every effort counts towards building a more sustainable and resilient world.

Evaluating Eco-Friendly Certifications & Labels

With the growing demand for eco-friendly and sustainable products, numerous certifications and labels have emerged to help consumers identify environmentally responsible choices. While these certifications can be helpful in guiding purchasing decisions, it's important to understand what each certification entails and how it aligns with your personal values and sustainability goals. Here are some key factors to consider when evaluating eco-friendly certifications and labels:

1. Transparency & Credibility:

- ❖ Look for certifications and labels that are backed by reputable organizations, third-party certifiers, or governmental agencies with established credibility and transparency.

- ❖ Research the certification process, standards, and criteria used to assess products, ensuring they are rigorous, science-based, and independently verified.

2. Comprehensive Coverage:

- ❖ Consider certifications that address multiple aspects of sustainability, including environmental, social, and economic factors, rather than focusing solely on one aspect.

- ❖ Look for certifications that consider a product's entire lifecycle, from raw material sourcing and production to distribution, use, and disposal, to ensure a holistic approach to sustainability.

3. Specific Environmental Criteria:

- ❖ Evaluate certifications based on specific environmental criteria relevant to the product category, such as energy efficiency, resource conservation, waste reduction, pollution prevention, and carbon emissions.

- ❖ Look for certifications that prioritize environmental impact reduction and promote sustainable practices throughout the supply chain, from sourcing of raw materials to manufacturing and distribution.

4. Independent Verification:

❖ Seek certifications that involve independent verification and auditing by accredited third-party organizations to ensure compliance with established standards and criteria.

❖ Look for certifications that require regular monitoring, reporting, and recertification to maintain compliance and accountability over time.

5. Recognized Standards & Labels:

❖ Familiarize yourself with recognized standards and labels in your region or industry, such as ENERGY STAR for energy-efficient appliances, USDA Organic for organic food products, or Forest Stewardship Council (FSC) for sustainably sourced wood and paper products.

❖ Research the meaning and significance of each certification and label to understand what they represent and how they align with your sustainability preferences.

6. Avoid Greenwashing:

- ❖ Be cautious of greenwashing, which refers to deceptive or misleading marketing claims that exaggerate the environmental benefits of a product without substantiating evidence.

- ❖ Look for certifications and labels that provide clear, transparent information about a product's environmental attributes and substantiate their claims with credible evidence.

7. Consider Local & Regional Certifications:

- ❖ Explore local and regional certifications and labels that may be more relevant to your geographic area or reflect specific environmental concerns and priorities.

- ❖ Support certifications that promote local sourcing, community engagement, and environmental stewardship within your region.

sunwave4u@gmail.com

8. Consumer Education & Awareness:

- ❖ Take the time to educate yourself about different eco-friendly certifications and labels, their meanings, and their relevance to your purchasing decisions.

- ❖ Stay informed about emerging trends, new certifications, and updates to existing standards to make informed choices that align with your values and sustainability goals.

In summary, evaluating eco-friendly certifications and labels requires careful consideration of factors such as transparency, credibility, comprehensive coverage, specific environmental criteria, independent verification, recognized standards, avoidance of greenwashing, consideration of local and regional certifications, and consumer education and awareness. By researching and understanding the meaning and significance of different certifications, consumers can make informed choices that support environmental sustainability and contribute to a healthier planet.

5.3 Conclusion:

Living by Science empowers you to take control of your life by making decisions based on evidence and reason. By applying the principles of science to your everyday choices, you can optimize your health, happiness, and well-being while also contributing to a healthier planet for future generations. Embrace the scientific approach to living, and unlock the full potential of your life.

About the Author:

Inderbir Singh is an explorer and researcher with a passion for exploring the sections of spirituality, religion, life, science, politics and personal growth. Drawing from a diverse background in philosophy, psychology, and theology, Inderbir Singh brings a nuanced perspective to the topics addressed in this book.

With a deep curiosity about the human experience and a commitment to fostering understanding and empathy, Inderbir Singh approaches each subject with an open mind and a willingness to explore complex issues from multiple angles. Their writing reflects a blend of scholarly inquiry, personal reflection, and practical wisdom, making the content accessible and engaging for readers of all backgrounds.

In addition to their work as a writer, Inderbir Singh is actively involved in community-building efforts, interfaith dialogue, and mindfulness practices. They believe in the power of storytelling, dialogue, and collective action to promote positive change and create a more harmonious and compassionate world.

sunwave4u@gmail.com

Inderbir Singh resides India, where they continue to write, research, and engage in meaningful conversations about spirituality, religion, science, politics and the human experience.

--Feel free to give any suggestion or feedback! ---

Thank You.

INDERBIR SINGH
Email: sunwave4u@gmail.com

www.ingramcontent.com/pod-product-compliance
Lightning Source LLC
Chambersburg PA
CBHW050105230526
45470CB00004B/1683